AfroBiotech Conference 2020

Online
26-28 October 2020

ISBN: 978-1-7138-2775-7

Printed from e-media with permission by:

Curran Associates, Inc.
57 Morehouse Lane
Red Hook, NY 12571

Some format issues inherent in the e-media version may also appear in this print version.

Printed with permission by Curran Associates, Inc. (2021)

For permission requests, please contact AIChE
at the address below.

AIChE
120 Wall Street, FL 23
New York, NY 10005-4020

Phone: (800) 242-4363
Fax: (203) 775-5177

www.aiche.org

Additional copies of this publication are available from:

Curran Associates, Inc.
57 Morehouse Lane
Red Hook, NY 12571 USA
Phone: 845-758-0400
Fax: 845-758-2633
Email: curran@proceedings.com
Web: www.proceedings.com

TABLE OF CONTENTS

MATHEMATICAL AND COMPUTATIONAL MODELING

METABOLIC ENGINEERING AND SYNTHETIC BIOLOGY

MICROBIOME AND IMMUNOENGINEERING

REGENERATIVE ENGINEERING

Deciphering and Engineering the Substrate Specificity of Protein-Modifying Enzymes

Conference: AfroBiotech Conference

Year: 2020

Proceeding: 2020 AfroBiotech Conference

Group: General Submissions

Session: Biological Chemistry and Nanotechnology

Authors:

Denard, C., University of Texas at Austin
Iverson, B. L., Univ. of Texas
Yaghi, R., University of Texas at Austin
Paresi, C., University of Texas at Austin
Yi, L., University of Texas at Austin

Protein-modifying enzymes (PMEs) are ubiquitous in biology, playing significant roles in initiating, regulating, and terminating cellular processes. The diversity and breadth of their substrate preference have been harnessed for a variety of applications in biotechnology and biomedicine. To optimize PMEs towards these applications, it is often necessary to engineer their catalytic properties and to profile their substrate specificities. In searching for desired activities in a large pool of variants, a high-throughput screening or selection system should ideally exhibit a broad operational range and a high dynamic range.

Here we describe YESS 2.0, a modular and customizable yeast endoplasmic sequestration screening system suitable for engineering and profiling the specificity of PMEs. By incorporating features to modulate gene transcription, as well as substrate and enzyme spatial sequestration within a versatile and seamless assembly method, YESS 2.0 achieves broad operational and dynamic range. To showcase YESS 2.0, we evolve a TEV protease variant (eTEV) with an 8-fold higher catalytic efficiency to obtain the fastest TEV protease variant to date. Second, we use YESS 2.0 coupled with NextGen Sequencing to profile the substrate specificity of insulin-degrading enzyme (IDE) and discover IDE substrates with increased activity compared to the prototypical insulin peptide. Lastly, we show for the first time that YESS 2.0 supports calcium-independent sortase-mediated ligations (SML) and confirm that residues directly C-terminal of the pentapeptide motif heavily influence the rate of SML. YESS 2.0 offers unmatched versatility in profiling and engineering the specificity of PMEs and should enable even more ambitious future undertakings.

Polycationic Nanogels for the Targeted Delivery of Micrornas

Conference: AfroBiotech Conference

Year: 2020

Proceeding: 2020 AfroBiotech Conference

Group: General Submissions

Session: Biological Chemistry and Nanotechnology

Authors:

Ward, D. M., University of Texas at Austin
Peppas, N. A., University of Texas at Austin

In the last few years, microRNA (miRNA) therapy has been proposed as a means to suppress tumorigenesis in cancer patients[1â€“3]. However, targeted delivery of miRNA remains a barrier for further implementation of this therapeutic strategy[4]. The tumor microenvironment differs markedly from the normal cell environment due to hypoxic regions around tumor cells. These regions of lowered oxygen and increased glycolysis result in local acidification[5]. In the present research, we have designed a novel targeted miRNA delivery system that utilizes pH-responsive polycationic nanogels. These polycationic carriers remain collapsed in basic and neutral conditions but are triggered by a change in local pH that would result in carrier swelling and associated miRNA release. Polycationic nanogels are synthesized via the controlled polymerization technique ARGET ATRP. Nanogels are then characterized and analyzed using Fourier transform infrared spectroscopy to characterize their chemical structure, dynamic light scattering to establish the carrier particles size distribution, and potentiometric titration to identify particle pH-responsiveness. Preliminary results of this work indicate that the nanogels exhibit favorable swelling behavior. These results suggest that this polycationic nanogel system is a promising system for miRNA delivery and potentially therapy.

This research was supported by grant No. R01-EB022025 from the National Institutes of Health.

References:

[1] Cy et al. (2018), *Open Chemistry,* 16(1), 1257-1267.

[2] Kota et al. (2009), *Cell,* 137(6), 1005-1017.

[3] Zeng et al. (2017), *Molecular Pharmaceutics,* 14(1), 81-92.

[4] Hosseinahli et al. (2018), *Journal of Cellular Physiology,* 233(8), 5574-5588.

[5] Boedtkjer and Pedersen (2020), *Annual Review of Physiology,* 82(1), 103-126.

Magnetic Propulsion and Manipulation of Achiral Swimmers in Polymer Fluids

Conference: AfroBiotech Conference

Year: 2020

Proceeding: 2020 AfroBiotech Conference

Group: General Submissions

Session: Biological Chemistry and Nanotechnology

Authors:

Quashie, D. Jr., Florida A&M University
Benhal, P., UNIVERSITY OF MARYLAND
Chen, Z., Southern University of Science and Technology
Wang, Z., Southern University of Science and Technology
Cheang, U. K., Southern University of Science and Technology
Ali, J., FAMU-FSU College of Engineering

Micro and nano scale swimmers, wirelessly controlled using magnetic fields, are envisioned to play a vital role in the future of targeted therapeutics. Thus far, design of these small scale devices has been guided by the Scallop theorem which states that, at low Reynoldâ€™s number, swimmers must exhibit non-time-symmetric motion. Common examples of artificial micro swimmer strategies include rotating rigid chiral helices, mimicking flagellated bacteria, and undulating flexible tails, reminiscent of spermatozoa locomotion. In addition to these methods of generating thrust, achiral objects that have no flexibility have been demonstrated to swim by meeting a minimum set of geometric requirements. These relatively simple swimmers can consist of linked particles or simple shapes that can be easily massively produced for potential application in biofluid environments. Here we characterize the swimming of mulit-bead achiral swimmers in different polymer fluids, demonstrating the effect of polymer chain length on swimming characteristics. We show the effect of bulk viscosity on the velocity profiles, step out frequency, and precession angle of the achiral swimmers. Our results show the ability to achieve controlled active propulsion of achiral swimmers in physiologically relevant viscosity ranges, providing evidence that these simple devices have potential for future in-vivo sensing and therapeutic applications.

Taxanes Sensitize Prostate Cancer Cells to TRAIL-Induced Apoptotic Synergy Via Endoplasmic Reticulum Stress

Conference: AfroBiotech Conference

Year: 2020

Proceeding: 2020 AfroBiotech Conference

Group: General Submissions

Session: Biological Chemistry and Nanotechnology

Authors:

Grayson, K., University of Michigan
Hope, J., Vanderbilt University
Wang, W., Vanderbilt University
Reinhart-King, C., Vanderbilt University
King, M., Vanderbilt University

Docetaxel (DTX) and cabazitaxel (CBZ) are guideline-chemotherapy treatments for metastatic castration-resistant prostate cancer (mCRPC), which comprises the majority of prostate cancer (PCa) deaths. TNF-related apoptosis inducing ligand (TRAIL) is an anti-cancer agent that is selectively cytotoxic to cancer cells; however, many human cancers are resistant to TRAIL.In this study, we sensitized androgen-independent and TRAIL-resistant PCa cells to TRAIL-mediated apoptosis via taxane therapy and examined the mechanism of sensitization. DU145 and PC3 cells displayed no significant reduction in cell viability when treated with soluble TRAIL, DTX, or CBZ alone indicating that both cell lines are resistant to TRAIL and taxanes individually. Taxane and TRAIL combination synergistically amplified apoptosis strongly suggesting that taxanes sensitize prostate cancer cells to TRAIL. A Jun N-terminal kinases (JNK) inhibitor inhibited apoptosis in treated cells and significantly reduced death receptor expression indicating JNK activation by ER stress sensitizes PCa cells to TRAIL-induced apoptosis by upregulating DR4/DR5 expression. In addition, suppression of C/EBP homologous protein (CHOP) reduced TRAIL sensitization in both cell lines indicating that ER stress-related apoptosis is mediated, in part by CHOP. Cytochrome *c* knockdown showed a significant decrease in sensitivity in PC3 cells, but not in Bax-deficient DU145 cells. A computational model was used to simulate apoptosis for cells treated with taxane and TRAIL therapy as demonstrated in *in vitro* experiments. Pretreatment with taxanes sensitized cells to apoptosis induced by TRAIL-mediated apoptosis, demonstrating that combining TRAIL with ER stress inducers is a promising therapy to reverse TRAIL resistance to treat mCRPC.

Oncogenesis and Aging By Isotopic Functionalizations of the Proteins and Nucleic Acids

Conference: AfroBiotech Conference

Year: 2020

Proceeding: 2020 AfroBiotech Conference

Group: General Submissions

Session: DNA Synthesis, Genome Design and Editing

Biosensors, which are used extensively in the field of research and development due to their easy availability, low cost, sensitivity and highly selective nature. So, they are used in the field of next generation medicine and point detection markers for the disease. The main types of biosensors are electrochemical, optical, piezoelectric and thermal biosensors.

Encephalitis is a disease caused by a mosquito borne virus. It was first recognized in India. Starting from a viral infection and later leading to inflammation of brain it is a rare disease. Carbon nanoparticles based electrochemical biosensors are used for the detection of the detection of this disease. In this technique, electrochemical immunosensors are used for the detection of encephalitis virus.

Innovative Approach to Minority Recruitment in a Genetic Study

Conference: AfroBiotech Conference

Year: 2020

Proceeding: 2020 AfroBiotech Conference

Group: General Submissions

Session: DNA Synthesis, Genome Design and Editing

Authors:

Johnson, A., Buffalo BioSciences
Cunningham, T., TruGenomix
Ewing, A., Genentech
Berro, T., Brigham and Women's Hospital
Linderman, M., Middlebury College
Green, R., Brigham and Women's Hospital

Background: African Americans remain underrepresented in genomics research, excluding African Americans from the benefits of medical discoveries and exacerbating health disparities (Bentley, 2017). One well-documented barrier to minority participation is mistrust in the medical system; however, some African Americans have noted a preference for racially and/or ethnically concordant researchers (George et al. 2014). In the pilot phase of the PeopleSeq Study, a proactive personal genome sequencing outcomes project, <1% of participants self-identified as African American or Black (Zoltick et al. 2019). To address the under-representation of African American participants, PeopleSeq obtained an NIH supplement to provide no-cost clinical genomic testing for individuals with African ancestry.

Purpose: To describe a multi-faceted recruitment approach built around educational events hosted by trusted African American leaders and community members who are engaged with and/or supportive of the research, combined with no-cost clinical genomic sequencing.

Methods: African American researchers and clinicians organized an educational event for community members to learn about genetics and proactive genome sequencing. Pre- and post-test genetic counseling and physician orders were coordinated to ensure a complete clinical experience outside of the PeopleSeq study.

Results: To date, 139 event participants expressed interest and 62 individuals have been invited via personal networks. Of those, 21 have elected to undergo sequencing. Recruitment is ongoing.

Conclusions: Participation of African American investigators, combined with community involvement, education sessions and providing no-cost testing can help increase minority representation in genomics research studies.

Oncogenesis and Aging By Isotopic Functionalizations of the Proteins and Nucleic Acids: Nonzero Nuclear Magnetic Moments By Quantum Criticality Alter Nucleophilicity and Electrophilicity

Conference: AfroBiotech Conference

Year: 2020

Proceeding: 2020 AfroBiotech Conference

Group: General Submissions

Session: DNA Synthesis, Genome Design and Editing

Authors:

Little, R. B., Stillman College

Although dynamics of telomeres during life expectancy of normal cells have been extensively studied, there are still some unresolved issues. For example, conditions required for telomere shortening leading to malignant transformations are not fully understood. In this work, we mass analyzed DNA of normal and cancer cells for comparing telomere isotopic compositions of white blood cells and cancer cells. We have found that characteristic telomere mass to charges of 1327 Da and 1672 Da cause differential mass distributions of about 1 Da for determining isotopic variations among normal cells relative to cancer cells. These isotopic differences are consistent with prior theory that replacing primordial, common isotopes of ^{1}H, ^{12}C, ^{14}N, ^{16}O, ^{24}Mg, ^{31}P and/or ^{32}S by nonprimordial, uncommon isotopes of ^{2}D, ^{13}C, ^{15}N, ^{17}O, ^{25}Mg and/or ^{33}S leads to altered enzymatic dynamics for modulating DNA and telomere codons towards transforming normal cells to cancer cells. Prior theory and current data are consistent also with a recently observed non-uniform methylation in DNA of cancer cells relative to more uniform methylation in DNA of normal cells. We observe further evidence of nonprimordial isotopic accelerations of acetylations, methylations, hydroxylations and aminations of nucleosides with alterations of phosphorylations of nucleotides for possibly explaining the induced mutations of DNA, RNA and proteins leading to cancer and more general alterations of DNA associated with aging. Different mass spectra of normal and cancer DNA may be reasoned by different functionalizations and isotopic enrichments as causing different motionally induced atomic and nucleotide orders by different nuclear magnetic moments.

Effective Engagement of Minority Students in an Introductory Biology Course through an Experimental Centric Pedagogy (ECP) That Utilizes an Analog Device to Measure Heart Rate

Conference: AfroBiotech Conference

Year: 2020

Proceeding: 2020 AfroBiotech Conference

Group: General Submissions

Session: Health and Medical Applications

Authors:

Owolabi, O., Morgan State University
Ariyibi, A., Morgan State University
Oni, A., Morgan State University
Ndirangu, C., Morgan State University
Bista, K., Morgan State University
Bhandari, A., Morgan State University
Ladeji-Osias, J., Morgan State University

With external funding, an evidence-based experimental centric pedagogy (ECP) is being implemented in Biology at a historically black university. This is the first of its kind where the ECP is being extended to several STEM disciplines after its successful implementation in Electrical Engineering to promote motivation as well as enhance academic achievement of minority students. During Spring and Summer semesters of 2020, ECP was developed and implemented in an introductory biology course. In Spring 2020, an analog measuring device was used to measure students heart rate, while in Summer due to COVID-19, in the remote class setting, monitored non chemical use, inexpensive home-based hands-on learning activities were developed to effectively engage the students. Research results show that the outcomes of the adaptation and use of the analog measuring device to detect and measure pulse rates at the tips of the fingers compared favorably to those obtained from the use of commercially available wrist monitors previously used for the experiments. Further, the various steps of the â€˜process of scienceâ€™ that the experimental design was meant to teach was equally achieved, and students were highly motivated as well as demonstrated interest in Biological related fields.

Snail Transctiption Factor in Prostate Cancer Cells Promotes Neurite Outgrowth

Conference: AfroBiotech Conference

Year: 2020

Proceeding: 2020 AfroBiotech Conference

Group: General Submissions

Session: Health and Medical Applications

Authors:

Edwards, G., Clark Atlanta University

Neurite outgrowth involves reciprocal signaling interactions between tumor cells and nerves where invading tumor cells have acquired the ability to respond to pro-invasive signals within the nerve environment. Neurite outgrowth could serve as a mechanism leading to invasion of cancer cells into the nerve sheath and subsequent metastasis. Snail transcription factor can promote migration and invasion of prostate cancer cells. We hypothesized that prostate cancer cell interaction with nerve cells will be mediated by Snail expression within prostate cancer cells. For this study we utilized various prostate cancer cell lines: C4-2 non-silencing (NS, control); C4-2 Snail shRNA, (stable Snail knockdown); LNCaP Neo (empty vector control) and LNCaP Snail (stably over-expressing Snail). Cancer cell adhesion and migration towards nerve cells (snF96.2 or NS20Y) was examined by co-culture assays. Conditioned media (CM) collected from C4-2 cells was cultured with nerve cells (PC-12 or NS20Y) for 48 hours followed by qualitative or quantitative neurite outgrowth assay. Our results showed that cancer cells expressing high levels of Snail (LNCaP Snail/C4-2 NS) displayed significantly higher migration adherence to nerve cells, compared to cells with lower levels of Snail (LNCaP Neo/C4-2 Snail shRNA). Additionally, LNCaP Snail or C4-2 NS (Snail-high) CM led to a higher neurite outgrowth compared to the LNCaP Neo or C4-2 Snail shRNA (Snail-low). In conclusion, Snail promotes migration and adhesion to nerve cells, as well as neurite outgrowth via secretion of soluble factors. Therefore, targeting cancer cell interaction with nerves may contribute to halting prostate cancer progression/metastasis.

Pre-Eclampsia: Revisiting Risk Factors and Challenging Current Approaches to Screening in All Settings

Conference: AfroBiotech Conference

Year: 2020

Proceeding: 2020 AfroBiotech Conference

Group: General Submissions

Session: Health and Medical Applications

Authors:

Lee, K., Brown University
Tripathi, A., Brown University
Brayboy, L., Brown University

The United Nations set a goal to reduce maternal mortality by three quarters between 1990 and 2015 in their Millennium Development Goals, which were not met. Most of these deaths are preventable, and one quarter of maternal deaths are due to pre-eclampsia and eclampsia. Prenatal screening is important for improving maternal health outcomes in general and is essential for the detection and management of pre-eclampsia. However, pre-eclampsia screening is solely based on maternal risk factors and has low (<5% in the U.S.) detection rates.

To address this global problem, risk factors should be analyzed at a more personalized level and studies should be performed in more low- and middle-income countries, non-European populations and at a community level. Additionally, access to technological screening tools must be more widespread.

This review describes the latest theories in pre-eclampsia etiology, the scientific basis behind established pre-eclampsia risk factors, biomarkers that have been identified as significant for pre-eclampsia, and technologies that aid in advancing pre-eclampsia screening worldwide. Researchers are challenged to think critically when performing and analyzing results of pre-eclampsia-related studies, especially when mistreating race as a risk factor, to encourage the development of improved alternatives to pre-eclampsia screening.

Biosensors in Encephalitis

Conference: AfroBiotech Conference

Year: 2020

Proceeding: 2020 AfroBiotech Conference

Group: General Submissions

Session: Health and Medical Applications

Authors:

Mishra, A. R. III, VELS Institute of Science Technology and Advanced Studies

Biosensors, which are used extensively in the field of research and development due to their easy availability, low cost, sensitivity and highly selective nature. So, they are used in the field of next generation medicine and point detection markers for the disease. The main types of biosensors are electrochemical, optical, piezoelectric and thermal biosensors.

Encephalitis is a disease caused by a mosquito borne virus. It was first recognized in India. Starting from a viral infection and later leading to inflammation of brain it is a rare disease. Carbon nanoparticles based electrochemical biosensors are used for the detection of the detection of this disease. In this technique, electrochemical immunosensors are used for the detection of encephalitis virus.

Investigating Liver Fibrosis in a Humanized Mice Model of Sickle Cell Disease

Conference: AfroBiotech Conference

Year: 2020

Proceeding: 2020 AfroBiotech Conference

Group: General Submissions

Session: Health and Medical Applications

Authors:

Michael, A., Georgia Institute of Technology
Platt, M., Georgia Institute of Technology & Emory University
Song, H., Georgia Institute of Technology & Emory University
Flowers-Steele, J., Georgia Institute of Technology

Sickle cell disease (SCD) is an inherited disorder that affects approximately 100,000 people in the United States and millions worldwide. Although severely understudied, frequency of acute and chronic liver pathologies has increased over the past decades due to the growing life expectancy of SCD patients. Typically, sterile inflammation in the liver promote activation of normally quiescent hepatic stellate cells (HSCs), which leads to excess production of cytokines and growth factors, such as transforming growth factor beta (TGF-Î²). The activation of HSCs leads to increased expression of profibrotic genes such as collagen, and alpha-smooth muscle actin (Î±-SMA). Increased production of these profibrotic genes lead to tissue fibrosis and unreversed fibrosis can progress to liver failure. Therefore, targeting TGF-Î² signaling is crucial to mitigate hepatic fibrosis. Previous studies have shown that cysteine cathepsins â€“ potent collagenases and elastases â€“ can modulate TGF-Î² protein levels. Based on these studies, we hypothesize that increased TGF-Î², through activation of HSCs, promotes liver fibrosis in SCD. To test this hypothesis, we utilized 3-month-old Townes mice liver and performed immunoblotting, zymography and immunostaining. Results showed increased bridging fibrosis, collagen content and Î±-SMA in sickle mice. Thereâ€™s increased phosphorylated Smad2/3 protein levels and increased active cathepsins in SS mice. Future work will delineate the signaling interplay of TGF-Î² signaling, HSCs activation/differentiation and the increased deposition of ECM proteins leading to liver fibrosis and lastly, we will determine whether potent collagenases such as cysteine cathepsins can modulate TGF-Î², thus reducing fibrosis in sickle cell disease.

Current Advancements in Biomedical Engineering

Conference: AfroBiotech Conference

Year: 2020

Proceeding: 2020 AfroBiotech Conference

Group: General Submissions

Session: Health and Medical Applications

Authors:

Bello, P., Obafemi Awolowo University

Biotechnology and biotech innovation is rapidly improving everyday as more fields are opening bringing huge potential for innovation and change. Artificial Intelligence and Machine Learning is one of the tools that is gaining wide acceptance in biotechnology and specifically in biomedical engineering. Once adequate and quality data is available, it makes it easier to study patterns and proffer solutions easier and faster.

Biosensor is another interesting development in medical technology. They are used to track blood pressure, pulse, breathing and body temperature. Biosensors can even help to track the amount of fluid in the lungs of heart failure patients and alert healthcare practitioners at the hospital, thereby improving healthcare delivery.

Another exciting development is 3D bioprinting. 3D printing also known as additive manufacturing has been around for some time and has helped in the evolution of 3D bioprinting. 3D bioprinting helps use living cells to create various human body parts such as heart valves, skin, cartilage, etc. for use in medicine. Bioprinting also has potentials to create drugs and pills as well as entire organs from a patients own body cells. It has also improved medical research and training.

Virtual and augmented reality also help to examine the 3d representation of the human body and various parts. It helps a lot in diagnosis. Quality healthcare helps to improve diagnosis, provides better treatment and quality solutions and biotechnology is really helping in achieving this objective.

Techno-Economic Assessment of Transforming Sorghum Bagasse into Bioethanol Fuel in Nigeria: (2) - Economic Analysis

Conference: AfroBiotech Conference

Year: 2020

Proceeding: 2020 AfroBiotech Conference

Group: General Submissions

Session: Mathematical and Computational Modeling

Authors:

Oyegoke, T., Federal University of Technology
Ajayi, O. O., Federal University of Technology
Onifade, K. R., Federal University of Technology

The Nigerian government's need to reduce its over-dependence on the oil and gas economy is gaining more significance as the benefit derived from this resource has gradually begun to suffer from several uncertainties. With the vast green resource, Nigeria is blessed with, this need is raising an avenue to establish a strong link between the downstream petroleum industry and agricultural activities. The Nigerian government has indicated a commitment to biofuels production from local feedstock where the emphasis has been given to bioethanol and biodiesel with the projected annual local market possibility of 5.04 billion and 900 million Liters, respectively. The study, therefore, looked into the investigation of the economic feasibility of establishing a bioethanol plant that would process sorghum bagasse into a bioethanol fuel that would complement the use of petrol. The sensitivity of selected economic parameters was also investigated. Results obtained for the profitability analysis of the project was able to showcase the benefit the project tends to offer alongside the identification of potential sensitive variables that have a strong potential of influencing the projectâ€™s financial benefit were presented with the use of parameters such internal rate of return, net present worth, and short payback period. Findings from the study would reveal that the potential investment benefits and would further gain investors' confidence by unfolding several uncertainties that investors would like to have some answers to about the setting up of a bioethanol plant in developing nations like Nigeria.

Minimizing Disease Spread on a Quarantined Cruise Ship: A Model of COVID-19 with Asymptomatic Infections

Conference: AfroBiotech Conference

Year: 2020

Proceeding: 2020 AfroBiotech Conference

Group: General Submissions

Session: Mathematical and Computational Modeling

Authors:

Rankin, N., Howard University
Dickenson, D., Howard University
Gurski, K., Howard University
Batista, B., Howard University
Kebe, M., Howard University

On February 5 the Japanese government ordered the passengers and crew on the Diamond Princess to start a two week quarantine after a former passenger tested positive for COVID-19. During the quarantine the virus spread rapidly throughout the ship. By February 20, there were 651 cases. We model this quarantine witha SEIR model including asymptomatic infections with differentiated shipboard roles for crew and passengers.The study includes the derivation of the basic reproduction number and simulation studies showing the effect of quarantine with COVID-19 or influenza on the total infection numbers. We show that quarantine on a ship with COVID-19 will lead to significant disease spread if asymptomatic infections are not identified. However, if the majority of the crew and passengers are immune or vaccinated to COVID-19, then quarantine would slow the spread. We also show that a disease similar to influenza, even with a ship with a fully susceptible crew and passengers, could be contained through quarantine measures.

Multiple Sites on Sars-Cov-2 Spike Protein Are Susceptible to Proteolysis By Cathepsins B, K, L, S, and V

Conference: AfroBiotech Conference

Year: 2020

Proceeding: 2020 AfroBiotech Conference

Group: General Submissions

Session: Mathematical and Computational Modeling

Authors:

Platt, M., Georgia Institute of Technology & Emory University
Bollavaram, K., Georgia Institute of Technology and Emory University
Leeman, T., Georgia Institute of Technology and Emory University
Lee, M., Georgia Institute of Technology and Emory University
Kulkarni, A., Georgia Institute of Technology and Emory University
Upshaw, S., Georgia Institute of Technology and Emory University
Yang, J., Georgia Institute of Technology and Emory University
Song, H., Georgia Institute of Technology and Emory University

SARS-CoV-2 is the coronavirus responsible for the COVID-19 pandemic that has killed more than 215,000 Americans. COVID-19 is also a health disparity with a disproportionate amount of infections and deaths occurring among African-American, Hispanic, and indigenous populations. Proteases are central to the infection process of SARS-CoV-2. Cleavage of the spike protein on the virusâ€™s capsid into S1/S2 subunits causes the conformational change that leads to membrane fusion and viral entry into the target cell. Since inhibition of one protease, even the dominant protease like TMPRSS2, may not be sufficient to block SARS-CoV-2 entry into cells, other proteases that may play an activating role and hydrolyze the spike protein must be identified. We identified amino acid sequences in all regions of spike protein, including the S1/S2 region critical for activation and viral entry, that are susceptible to cleavage by furin and cathepsins B, K, L, S, and V using PACMANS, a computational platform that identifies and ranks preferred sites of proteolytic cleavage on substrates, and verified with molecular docking analysis to determine if binding of these proteases can occur on the spike protein that were identified as possible cleavage sites. Together, this study highlights cathepsins B, K, L, S, and V for consideration in SARS-CoV-2 infection and presents methodologies by which other proteases can be screened to determine a role in viral entry or rationale for inhibition in patients with inflammatory cardiovascular or pulmonary disease.

Modeling the Spatial Distribution of Collagen in Mesangial Fibrosis during Diabetic Kidney Disease

Conference: AfroBiotech Conference

Year: 2020

Proceeding: 2020 AfroBiotech Conference

Group: General Submissions

Session: Mathematical and Computational Modeling

Authors:

Thomas, H. Y., Oklahoma State University
Ford Versypt, A. N., Massachusetts Institute of Technology

In the U.S. alone over 250,000 people use dialysis or have received a kidney transplant due to diabetic kidney failure. Although we have come a long way in the treatment of diabetes, kidney failure due to diabetic kidney damage is still prevalent, and the need for increasing our understanding of kidney damage to enable the development of better treatment methods is ever present. Thus the goal of this research is to develop computational models to better understand the kidney damage that occurs due to diabetic kidney disease. In the kidney glomerulus, the filtration unit of the kidney, lies a network of capillaries that are surrounded by interstitial tissue called the mesangium. In health, the mesangium acts as a support for the capillaries; however, during diabetic kidney disease, the mesangium undergoes excess deposition of collagen leading to irreparable damage to the environment around it. This damage occurs in stages that have marked spatial distribution patterns. Initially, the deposition is just diffuse collagen deposition which progresses onto nodular formations then segmental collagen deposition until the final stage of globular collagen deposition. Our goal is to understand the factors that affect the spatial distribution of collagen in mesangial fibrosis. Thus, we have developed an agent based model of mesangial fibrosis, using fundamental biological principles of collagen fiber growth, to enable the prediction of spatial deposition patterns and in doing so develop models to predict the progression of diabetic kidney damage.

Data-Driven Process Modeling for Prediction of Outcomes for Cardiac Differentiation Experiments

Conference: AfroBiotech Conference

Year: 2020

Proceeding: 2020 AfroBiotech Conference

Group: General Submissions

Session: Mathematical and Computational Modeling

Authors:

Williams, B., Auburn University
Halloin, C., Hannover Medical School
Loebel, W., Hannover Medical School
Finklea, F., Auburn University
Lipke, E., Auburn University
Zweigerdt, R., Hannover Medical School
Cremaschi, S., Auburn University

Cardiovascular diseases (CVD) are the leading cause of death worldwide, meaning that more people die each year of CVD than of any other cause. These diseases can lead to heart attacks, which can result in the loss of more than one billion heart cells and lead to congestive heart failure. Engineered heart tissue produced by differentiation of human induced pluripotent stem cells may provide an encompassing treatment for heart failure due to CVD. However, significant difficulties exist in producing the large number of cardiomyocytes needed for therapeutic purposes through differentiation protocols. Data-driven modeling with machine learning techniques represents a potential tool for elucidating the relationship between experimental conditions and the outcomes for cardiac differentiation experiments. Using data from previously conducted cardiac differentiation experiments, we have developed data-driven modeling methods along with feature selection for determining which experimental conditions are most influential on and predictive of the final cardiomyocyte content of a differentiation experiment. With those identified conditions, we were able to build classification models that can predict whether an experiment will have a sufficient cardiomyocyte content to continue with the experiment on the seventh (out of 10) day of the differentiation with a 90% accuracy. This early failure prediction will provide cost and time savings, as each day the differentiation continues requires significant resources.

Predicting the Effects of Engineering T Cells Using Systems Biology Modeling

Conference: AfroBiotech Conference

Year: 2020

Proceeding: 2020 AfroBiotech Conference

Group: General Submissions

Session: Mathematical and Computational Modeling

Authors:

Finley, S. D., University of Southern California

Systems biology approaches, including computational models, provide a framework to test biological hypotheses and optimize effective therapeutic strategies to treat human diseases. In this talk, I present recent work in modeling signaling in cancer-targeting immune cells, with a focus on CAR T cells. Chimeric antigen receptors (CARs) are comprised of a variety of different activating domains and co-stimulatory domains that initiate signaling required for T cell activation. However, there is a lack of understanding of the mechanisms by which activation occurs. We aim to fill this knowledge gap by applying mathematical modeling to investigate how CAR structure influences downstream T cell signaling and develop new hypotheses for the optimal design of CAR-engineered T cell systems. Our work addresses questions regarding the role of signaling domains and kinase activity in the activation of engineered T cells and can be used to guide CAR T cell design.

Genome-Scale Metabolic Modeling Allows for Greater Understanding of Gene Function in Cryptosporidium Parvum

Conference: AfroBiotech Conference

Year: 2020

Proceeding: 2020 AfroBiotech Conference

Group: General Submissions

Session: Mathematical and Computational Modeling

Authors:

Olusanya, O., Howard University
Carey, M. A., University of Virginia

Advances in technology have allowed for an increase in our genomic knowledge, stored in a wide range of databases. However, many of these sequenced genes are annotated incorrectly or not at all, so we remain uncertain of their exact function in various organisms. To solve this problem, we use genome-scale metabolic models to identify essential enzymes and metabolites, find their functional homologs in model organisms like *E. coli*, and engineer growth conditions for the model organisms that allow us to find the genes responsible for the enzymes. The genomic library of the organism of interest is transformed into E. coli recipient strains, and the bacteria that survive in the growth conditions are known to contain the genes of interest. These genes are then sequenced and added to the model.

We are applying this novel approach to *Cryptosporidium parvum*, a parasite that causes cryptosporidiosis by infects the epithelial cells of mammalian intestinal tracts. This infection causes diarrhea in humans which can result in death in children who are infected. Currently there is one treatment available for infected adults and none available for children. Parasites like *Cryptosporidium parvum* have had their genomes sequenced, but much of the genome is unannotated and, thus, there are many gaps in our knowledge of their metabolism that we can fill using this metabolic modeling system. The information collected from these models can be used to create drugs and treatments that target these metabolic pathways.

RNA-Seq: A Transcriptomics Approach to Understanding Protein Production in CHO Cells

Conference: AfroBiotech Conference

Year: 2020

Proceeding: 2020 AfroBiotech Conference

Group: General Submissions

Session: Metabolic Engineering and Synthetic Biology

Authors:

Scott, D., Delaware State University

Recombinant therapeutic proteins are widely used to treat life-threatening diseases including cancer. Chinese hamster ovary (CHO) cells have been used for over 25 years in the biopharmaceutical industry to produce recombinant therapeutics, where CHO cells produce almost one in three of the manufactured recombinant therapeutics. CHO cell lines are the preferred host for protein expression due to their ability to produce human-compatible protein glycosylation, mediate protein folding, ease of culturing in large-scale bioreactors and protein expression stability. However, several limitations exist in CHO cell bioprocessing, including low protein yields and genome instability, resulting in high market prices of the biopharmaceuticals. Here, we discuss using RNA transcriptomics on cells grown under varying industry relevant conditions to discover genes that are important to the protein production process by increasing stabilization. This knowledge has the potential to decrease the costs of medicines through the minimization of waste products in the CHO culturing procedure.

In silico Prediction of Metabolic Targets for 3-Hydroxypropanoate Production in E. coli

Conference: AfroBiotech Conference

Year: 2020

Proceeding: 2020 AfroBiotech Conference

Group: General Submissions

Session: Metabolic Engineering and Synthetic Biology

Authors:

Chaves, G., Federal University of S�o Carlos
Silva, A., Federal University of Sao Carlos

3-hydroxypropionate (3-HPP) has great potential to serve as a building block for the chemical industry as its bifunctional character and high reactivity make it suitable for replacing petroleum-derived chemicals. Since the chemical processes proposed for 3-HPP synthesis are highly toxic and environmentally harmful, this compound is not produced on an industrial scale nowadays. Therefore, fermentative bioprocesses are promising alternatives for producing 3-HPP. Hence, this work aimed to identify non-intuitive reactions that can be engineered for optimizing 3-HPP production in *E. coli* by the Î^2-alanine pathway. First, the corresponding reactions for 3-HPP biosynthesis were introduced into a genomic-scale metabolic model of *E. coli* K-12 MG1655 (*i*ML1515). Subsequently, using the software Optflux, *in silico* 3-HPP production was optimized with an evolutionary algorithm that searched for suitable flux values for the reactions in the model, by under/overexpressing them to enhance 3-HPP formation using xylose or glucose as carbon sources. Results showed that alanine racemase overexpression and, L-valine transaminase and L-alanine transaminase underexpression were the main targets. These modifications were found to increase L-glutamate while decreasing L-alanine pools, thus favoring L-aspartate aminotransferase and Î^2-alanine: pyruvate transaminase reactions towards 3-HPP production. These results will be further assessed *in vivo* in a 3-HPP production strain.

Seeking Safe Stool: Experiences from an Non-Profit Stool Bank

Conference: AfroBiotech Conference

Year: 2020

Proceeding: 2020 AfroBiotech Conference

Group: General Submissions

Session: Microbiome and Immunoengineering

Authors:

Osman, M., OpenBiome

Fecal microbiota transplantation (FMT) is an established therapy for the treatment of recurrent *C. difficile* infection (rCDI). FMT has already advanced significantly from a DIY therapy using minimally screened individual donors to a highly standardized process, using universal donors subject to rigorous screening.

As an example of this transformation, OpenBiome, the world's largest non-profit public stool bank, has delivered over 55,000 treatments for rCDI to over 1,300 hospitals in 7 countries, with less than 3% of prospective donors passing the 200-point clinical assessment and 30-item laboratory screening panel required for enrollment. OpenBiome also support clinical trials of FMT in other disease indications including inflammatory bowel disease, antibiotic resistant infections, and autoimmune diseases.

Seeking Safe Stool: Experiences from an International Public Stool Bank

Conference: AfroBiotech Conference

Year: 2020

Proceeding: 2020 AfroBiotech Conference

Group: General Submissions

Session: Microbiome and Immunoengineering

Authors:

Osman, M., OpenBiome

Advances in technology have allowed for an increase in our genomic knowledge, stored in a wide range of databases. However, many of these sequenced genes are annotated incorrectly or not at all, so we remain uncertain of their exact function in various organisms. To solve this problem, we use genome-scale metabolic models to identify essential enzymes and metabolites, find their functional homologs in model organisms like *E. coli*, and engineer growth conditions for the model organisms that allow us to find the genes responsible for the enzymes. The genomic library of the organism of interest is transformed into E. coli recipient strains, and the bacteria that survive in the growth conditions are known to contain the genes of interest. These genes are then sequenced and added to the model.

We are applying this novel approach to *Cryptosporidium parvum,* a parasite that causes cryptosporidiosis by infects the epithelial cells of mammalian intestinal tracts. This infection causes diarrhea in humans which can result in death in children who are infected. Currently there is one treatment available for infected adults and none available for children. Parasites like *Cryptosporidium parvum* have had their genomes sequenced, but much of the genome is unannotated and, thus, there are many gaps in our knowledge of their metabolism that we can fill using this metabolic modeling system. The information collected from these models can be used to create drugs and treatments that target these metabolic pathways.

Design of a Novel Poly(dimethylsiloxane)-Based Microbial Nanoculture System

Conference: AfroBiotech Conference

Year: 2020

Proceeding: 2020 AfroBiotech Conference

Group: General Submissions

Session: Microbiome and Immunoengineering

Authors:

Usman, H., University of Illinois at Chicago

Organosilanes contain hydrocarbon-like backbones, allowing them to react with silicone-based agents in the presence of a catalyst and polymerize into membranes with tunable transport and mechanical properties. Owing to their high hydrophobicity, Poly(dimethylsiloxane) (PDMS) membranes, and more particularly, SylgardÂ® 184, have been used for applications including drug delivery, gas separation, and microfluidics fabrication. However, the undefined composition of the material and its ability to leach out uncured oligomers make its functionalization and usage challenging for many biological applications. This article presents the design of a novel culture system generated using PDMS-based membranes to study microbial dynamics. The microbial culture system that is referred to as â€œnanocultureâ€☐ serves to encapsulate and grow microbes in semipermeable membranes. The mechanical properties of the membranes are investigated through osmotic annealing to determine the ability of the nanocultures to withstand high shear stress similar to environmental conditions while maintaining transport properties essential to microbial communication and growth. The present study lays the foundation for a novel microbial culture system that would enable the cultivation of microorganisms in environments other than laboratory conditions.

Screening for Thermostable Extracellular Amylases from Bacterial Isolates

Conference: AfroBiotech Conference

Year: 2020

Proceeding: 2020 AfroBiotech Conference

Group: General Submissions

Session: Microbiome and Immunoengineering

Authors:

Ullah, I., The university of Agriculture peshawar

Synthetic small-diameter vascular grafts are often a treatment for cardiovascular disease; yet failure due to thrombosis remains a problem. *In situ* endothelialization of graft surfaces can mitigate thrombosis. Poly(vinyl alcohol) hydrogel (PVA) is a hemocompatible biomaterial for grafting, however, its bioinert properties prevent endothelialization. Mixing extracellular matrix (ECM)-mimicking biomolecules into PVA networks increased endothelial cell (EC) attachment. Yet, platelet attachment was concomitantly increased, indicating an alteration of the materialâ€™s antithrombotic properties. Herein, we hypothesized that covalent attachment of ECM-mimicking biomolecules to PVA would specifically bind ECs without compromising hemocompatibility. We covalently attached a collagen-mimicking peptide, GFPGER, to surfaces of planar and tubular PVA. The surfaces of crosslinked PVA were activated with carbonyldiimidizole before immersion into GFPGER peptide solutions (0, 15, 30, 60 and 120Âµg/mL). To characterize the surface properties of the modified materials, we measured contact angle and performed XPS elemental analyses. EC attachment was quantified with a DNA assay and qualitatively examined using immunohistochemistry by staining for VE-cadherin, actin, and nuclei. We quantified hemocompatibility with *in vitro* platelet attachment and time to fibrin clot formation and *ex vivo*platelet attachment and fibrin formation in an established non-human primate whole blood shunt model. EC attachment to GFPGER-modified PVA was increased. *In vitro* and *ex vivo*hemocompatibility results showed a reduction in platelet attachment and fibrin formation onto GFPGER-modified PVA. Our results confirmed that covalently modifying PVA with GFPGER promotes EC attachment while maintaining hemocompatibility. This work is a significant step toward development of hemocompatible devices for treatment of cardiovascular disease.

3D in Vitro Neuron on a Chip for Probing Calcium Mechanostimulation

Conference: AfroBiotech Conference

Year: 2020

Proceeding: 2020 AfroBiotech Conference

Group: General Submissions

Session: Regenerative Engineering

Authors:

Bobo, J., CMU
LeDuc, P. R., Carnegie Mellon University

Creating an environment that mimics the dimensionality and complexity of tissue in the nervous system that can respond to mechanical stimuli to determine injury, responses, and rehabilitation timelines could provide an advance in physiologically relevant systems. One exciting approach toward this is to focus on the mechanobiology perspective merged with a 3D nerve integrated tissue on a chip system, which could potentially aid in the development of protocols and severity assessment of injuries to the brain. Current in vitro 2D systems, while having made great progress over the last decade, are limited by 2D cell-extracellular matrix interactions, which restricts the establishment of a microenvironment comparable to that of 3D neuronal network in the brain. Here we describe an experimental 3D biomaterial in which variational loads and rates are applied to the lab on a chip. Biomaterials, such as collagen, provide a biocompatible, non-toxic ECM to model the brain induced with trauma. We labelled the cells with Fluo-4-AM to examine their calcium response under mechanical stimulation. We imposed mechanical stimulation synchronized with image capturing to examine the cellular response. We observed an increase in calcium during stimulation which was related to strain rate.

Boo

Conference: AfroBiotech Conference

Year: 2020

Proceeding: 2020 AfroBiotech Conference

Group: General Submissions

Session: Regenerative Engineering

Authors:

Boo, B., Boo

Biotechnology and biotech innovation is rapidly improving everyday as more fields are opening bringing huge potential for innovation and change. Artificial Intelligence and Machine Learning is one of the tools that is gaining wide acceptance in biotechnology and specifically in biomedical engineering. Once adequate and quality data is available, it makes it easier to study patterns and proffer solutions easier and faster.

Biosensor is another interesting development in medical technology. They are used to track blood pressure, pulse, breathing and body temperature. Biosensors can even help to track the amount of fluid in the lungs of heart failure patients and alert healthcare practitioners at the hospital, thereby improving healthcare delivery.

Another exciting development is 3D bioprinting. 3D printing also known as additive manufacturing has been around for some time and has helped in the evolution of 3D bioprinting. 3D bioprinting helps use living cells to create various human body parts such as heart valves, skin, cartilage, etc. for use in medicine. Bioprinting also has potentials to create drugs and pills as well as entire organs from a patients own body cells. It has also improved medical research and training.

Virtual and augmented reality also help to examine the 3d representation of the human body and various parts. It helps a lot in diagnosis. Quality healthcare helps to improve diagnosis, provides better treatment and quality solutions and biotechnology is really helping in achieving this objective.

Cathepsin K, L, and S Differentially Regulate Arterial Compliance in Mouse Carotid Arteries

Conference: AfroBiotech Conference

Year: 2020

Proceeding: 2020 AfroBiotech Conference

Group: General Submissions

Session: Regenerative Engineering

Authors:

Omojola, V., Georgia Institute of Technology

Cysteine cathepsins are potent lysosomal proteases implicated in cardiovascular diseases such as atherosclerosis, sickle cell disease, and aneurysms. Proteolytic action by cathepsins degrades various substrates in the vasculature, most notably ECM, whose structure and integrity determine the mechanical behavior of arteries. Small animal models utilizing cathepsin knockout and atherogenic diets have been used to study changes in biomechanics in cardiovascular disease, but knowledge of the impacts of cathepsin knockout on non-atherosclerotic arterial mechanics are scarce. We aim to examine arterial mechanics in several cathepsin knockout lines ($CatK^{-/-}$, $CatL^{-/-}$ and CatS-/-) and controls (C57/Bl6, apolipoprotein $E^{-/-}$). Common carotid arteries of three month-old mice were isolated and underwent biaxial mechanical testing and opening angle tests. Pressure-diameter data revealed $CatL^{-/-}$ arteries were smaller in caliber compared to $CatK^{-/-}$, $CatS^{-/-}$, and $ApoE^{-/-}$ controls and were less compliant than $ApoE^{-/-}$ and $CatS^{-/-}$ arteries at lower pressures, where elastin governs the mechanical response. $CatK^{-/-}$ arteries had showed increased in vivo strains compared to $CatL^{-/-}$and $CatS^{-/-}$ arteries. $CatL^{-/-}$ arteries were less compliant than $ApoE^{-/-}$ and $CatS^{-/-}$ arteries pressurized to sub-diastolic pressures,. Taken together, the data suggests that the loss of the potent collagenase cathepsin K increases axial and circumferential arterial compliance, while knockout of the elastase cathepsin L decreased circumferential arterial compliance, and knockout of the elastase cathepsin S showed no impact on arterial mechanics.

Maximizing Force Production from hiPSC-Derived Engineered Cardiac Tissues By Stimulating Calcium Handling

Conference: AfroBiotech Conference

Year: 2020

Proceeding: 2020 AfroBiotech Conference

Group: General Submissions

Session: Regenerative Engineering

Authors:

Minor, A., Brown University
Coulombe, K. L. K., Brown University

Human induced pluripotent stem cell derived cardiomyocytes (hiPSC-CMs) have profound utility in generating functional human engineered cardiac tissue (ECTs) for heart repair. However, the field at large is concerned about the relative immaturity of these hiPSC-CMs as we aim to develop clinically relevant models for regenerative therapy and drug testing. Addressing the calcium handling properties and metabolic maturity will improve excitation-contraction (E-C) coupling and E-C metabolism coupling in hiPSC-CMs to facilitate functional integration *in vivo*. Herein, we develop a novel calcium (Ca^{2+}) conditioning protocol that maintains ECTs in physiological levels of Ca^{2+} and assess contractility in increasing calcium environments. Lactate selection served as a method to shift the metabolic profile of hiPSC-CMs to evaluate the role of metabolism on Ca^{2+} sensitivity. We observe two-fold greater contractility in the calcium conditioning group with lower stiffness. Interestingly, the force-calcium relationship reveals higher calcium sensitivity in lactate conditioned tissues suggesting that metabolic maturation alters mitochondrial Ca^{2+} buffering capacity. We find that calcium sensitivity and amplitude of contraction are not coupled, as lactate conditioned tissues produce force comparable to that of controls in high calcium environments. An upregulation of calcium handling proteins further contributes to the greater Ca^{2+} sensitivity in lactate conditioned hiPSC-CMs. Our findings support the use of physiological $Ca^{2+}$to enhance the functional maturation of E-C coupling in hiPSC-CMs and demonstrate that metabolic changes induced by lactate conditioning significantly alters cell sensitivity to external calcium. These conditioning methods may further the development of 3D human engineered cardiac tissue as a potential regenerative therapy.

Covalent Modification of Poly(vinyl alcohol) Hydrogels with a Collagen-Mimicking Peptide Encourages Endothelialization While Minimizing Thrombosis

Conference: AfroBiotech Conference

Year: 2020

Proceeding: 2020 AfroBiotech Conference

Group: General Submissions

Session: Regenerative Engineering

Authors:

Bates, N., Oregon Health & Science University
Heidenreich, H., Oregon Health & Science University
Fallon, M., Oregon Health & Science University
Yao, Y., University of Waterloo
Yim, E., University of Waterloo
Hinds, M., Oregon Health & Science University
Anderson, D., Oregon Health & Science University

Synthetic small-diameter vascular grafts are often a treatment for cardiovascular disease; yet failure due to thrombosis remains a problem. *In situ* endothelialization of graft surfaces can mitigate thrombosis. Poly(vinyl alcohol) hydrogel (PVA) is a hemocompatible biomaterial for grafting, however, its bioinert properties prevent endothelialization. Mixing extracellular matrix (ECM)-mimicking biomolecules into PVA networks increased endothelial cell (EC) attachment. Yet, platelet attachment was concomitantly increased, indicating an alteration of the materialâ€™s antithrombotic properties. Herein, we hypothesized that covalent attachment of ECM-mimicking biomolecules to PVA would specifically bind ECs without compromising hemocompatibility. We covalently attached a collagen-mimicking peptide, GFPGER, to surfaces of planar and tubular PVA. The surfaces of crosslinked PVA were activated with carbonyldiimidizole before immersion into GFPGER peptide solutions (0, 15, 30, 60 and 120Âµg/mL). To characterize the surface properties of the modified materials, we measured contact angle and performed XPS elemental analyses. EC attachment was quantified with a DNA assay and qualitatively examined using immunohistochemistry by staining for VE-cadherin, actin, and nuclei. We quantified hemocompatibility with *in vitro* platelet attachment and time to fibrin clot formation and *ex vivo*platelet attachment and fibrin formation in an established non-human primate whole blood shunt model. EC attachment to GFPGER-modified PVA was increased. *In vitro* and *ex vivo*hemocompatibility results showed a reduction in platelet attachment and fibrin formation onto GFPGER-modified PVA. Our results confirmed that covalently modifying PVA with GFPGER promotes EC attachment while maintaining hemocompatibility. This work is a significant step toward development of hemocompatible devices for treatment of cardiovascular disease.

In Vitro Cytotoxicity Studies of Novel Mesoporous Inorganic Polymers

Conference: AfroBiotech Conference

Year: 2020

Proceeding: 2020 AfroBiotech Conference

Group: General Submissions

Session: Regenerative Engineering

Authors:

Akono, A. T., Northwestern University
Wang, X., Northwestern University
Duan, C., Northwestern University
Ameer, G., Northwestern University

Bone is one of the most frequently transplanted tissues with 8 million surgical procedures performed annually in the United States alone. The leading practice is the use of autologous graft. However, given the high demand for bone grafts and the short supply of autologous graft, new biomaterial alternatives are needed that are bio-inductive, bio-conductive, and biocompatible. We focus on a new breed of potential bone grafts: Inorganic polymers. Inorganic polymers are X-ray amorphous polymers that exhibit a hierarchical architecture similar to that of cortical bone: nanogranular and mesoporous. We examine the bio-inductivity via biomineralization experiments. We evaluate the bio-conductivity by studying the cell-wall interactions of mouse fibroblast cells. Our preliminary results have shown that inorganic polymers are bio-inductive. Calcium phosphate crystals were observed after immersion of inorganic polymers in simulated body fluid solution. Preliminary results have also shown that inorganic polymers are resorbable as a decrease in weight was observed after accelerated biodegradation experiments. We will show the results of additional biomineralization and biodegradation studies. Moreover, we will examine the potential of inorganic polymers to promote fibroblast cells' adhesion and migration. These results are important to yield novel bone scaffolds for bone tissue regenerative engineering.

Exploring the Roles of Insulin and Insulinoma Cells in Keratinocyte Migration

Conference: AfroBiotech Conference

Year: 2020

Proceeding: 2020 AfroBiotech Conference

Group: General Submissions

Session: Regenerative Engineering

Authors:

Aubdoollah, Z., Rutgers University
Soni, S., Drexel University
Iturbide, E., Rutgers University
Olabisi, R., University of California

We previously demonstrated accelerated wound healing in an *in vitro* wound healing model and in genetically diabetic mice using hydrogel-encapsulated rat insulinoma cells (RIN-m) [1]. However, the results were obtained with an insulinoma cell line. Since cancer cells can promote growth and resist apoptosis [2], we examined keratinocyte response to RIN-m derived cell lines that delivered insulin (RIN-5F) or somatostatin (RIN-14B), an insulin inhibitor, to confirm the role of insulin in the accelerated wound healing.

Commercial insulin or conditioned media collected at weekly time points from monolayer or polyethylene glycol diacrylate encapsulated RIN-m, RIN-5F, or RIN-14B cells were evaluated for insulin levels and ability to increase keratinocyte migration in an *in vitro* scratch model of wound healing.

Encapsulated RIN-5F cells released the most insulin at 8.9Â±3.633 ng/mL, RIN-14B cells were no different than media controls. Commercial insulin and RIN-5F accelerated keratinocyte migration similarly. There was no difference between media controls and RIN-14B conditioned media.

In light of no accelerated *in vitro* wound healing response when using insulinoma derived cells lacking insulin, combined with the accelerated response when using commercial or insulinoma derived insulin, we conclude that the insulin is responsible for the accelerated healing and not some unidentified characteristic of cancer cells.

1. Aijaz, Ayesha, et al. Tissue engineering part A 21.21-22 (2015): 2723-2732.
2. Arwert, E.N., et al., Nature Reviews Cancer, 2012. 12(3): p. 170-180.

AUTHOR INDEX